YOUR KNOWLEDGE HAS VALUE

- We will publish your bachelor's and
  master's thesis, essays and papers

- Your own eBook and book -
  sold worldwide in all relevant shops

- Earn money with each sale

Upload your text at www.GRIN.com
and publish for free

Bibliographic information published by the German National Library:

The German National Library lists this publication in the National Bibliography; detailed bibliographic data are available on the Internet at http://dnb.dnb.de .

Imprint:

Copyright © 2012 GRIN Verlag, Open Publishing GmbH
Print and binding: Books on Demand GmbH, Norderstedt Germany
ISBN: 978-3-668-03742-7

A. Al-Ahaideb, N. Arain, S. Hashish, M. Dahlawi, Am. Elhossainy, As. Elhossainy, Y. Al Najashi, Y. Al Doghaither

Aging with Grace. Malnutrition in Elderly

GRIN Publishing

Aging with Grace:

Malnutrition in the Geriatric Population

Prepared By:

A. Al-Ahaideb, N. Arain, S. Hashish, M. Dahlawi, Amr Elhossainy, Asem Elhossainy, Y. Al Najashi, and Y. Al Doghaither

Aging With Grace: Malnutrition in Elderly

Abdallah Al-Ahaideb, Saeb Hashish, Nasir Nisar Arain, Asem El-Hossainy; Amr Elhossainy, Mohammad Dahlawi;

Yazeed Al-Dugaither, and Yousef Al-Najashi

Medical Students, Alfaisal University

Wedensday, January 04, 2012

College of Medicine, Alfaisal University

Supervisor: M. H. Rajab, PhD. MPH.

Abstract:

Introduction:

Malnutrition is very common in elderly, and as average life expectancy is increasing with correlate decrease in fertility and mortality rates worldwide, this change will shift the epidemiology towards more geriatric populations (>60 yrs old). There are different tools to evaluate nutritional status; however, the most widely used and adapted tool is the Mini Nutritional Assessment (MNA) Tool.

Goal:

The aim of this report study is to draw a random sample from elderly patients and evaluate their nutritional status retrospectively 3 months before their admission. It aims to link malnutrition with coexistence of relevant diseases and a literature review will be conducted accordingly. It also aims at increasing awareness amongst elderly people and youth about successful aging, and conducting a field study to an Elderly Care Centre to investigate its roles and duties in helping the geriatric community in Riyadh.

Methods:

This study was carried out on elderly patients (17 cases; > 60 yrs old), residing at in-patients wards in KFSH & RC in Riyadh, Saudi Arabia. In this study, data for 17 elderly patients was collected using a structured Mini Nutritional Assessment (MNA) Tool. Literature review from different international journals was conducted using Pubmed, Science Direct, and other research engines. To increase awareness, a pamphlet is intended to be emailed through Alfaisal University Electronic Email Service to educate youths and their relatives about the importance of successful aging. A field-study to an Elderly Care Centre in Nakheel District, Riyadh, was carried out by interviewing several personnel working for the house.

Results:
A total of 20 patients (male/female: 1:1) were included in the study. MNA showed poor nutritional status in 85% of patients (30% at risk (MNA score <=23), 55% malnutrition (MNA score<17)).The prevalence of malnutrition in females is significantly higher than that in males (P<0.05).

Conclusion:

The study confirms that there is co-existence between malnutrition and other diseases, and that malnutrition is a common problem among elderly. Though, due to the small sample size, this cannot be generalized for all population. We also encourage awareness campaigns on how to age successfully and have better health to lessen the burdens and economical costs that can be prevented by sheer knowledge and practice.

Introduction:

Malnutrition, which is the deficiency of energy, protein, and/or other nutrients such as vitamins, occurs with a high prevalence in the elderly and leads to significant increase in morbidity and mortality in a growing segment of the population of the world that's becoming a burden on the health care system.

Reported prevalence of malnutrition across the world varies greatly from 0-65% due to differences in methodologies employed and the heterogeneity of the aging population. In a multinational study spanning 12 countries and using 24 data sets, the prevalence of malnutrition was found to be 22.8% on average, with the setting being a source of variation (50.5% in rehabilitation, 38.7% in hospitalization, 13.8% in nursing homes, 5.8% in community). 46.2% were found to be "at risk" of malnutrition [1]. In the UK, the prevalence is reported to be 25%-35% in age groups between 60 and 80 yrs and 35% in adults over 80 [2]. In a study in China, prevalence was found to be 1.71% and 19.6% were "at risk" of malnutrition [3]. In the Czech Republic, malnutrition is reported to affect 20-70% of elderly patients in a hospital setting [4]. In Belgium, malnutrition among the elderly population is estimated to be 33% with 43% at risk [28].

Malnutrition contributes significantly to morbidity and mortality rates in elderly patients. A 10 year longitudinal study, in which 26.5% became weak or sick and 19.1% died, showed an increase in morbidity in malnourished patients. Subjects on a lower or higher calorie diet than the recommended (RDA) and those with a low protein intake were exposed to more weakness, sickness, or death than their counterparts [5]. Studies on elderly surgical patients suggest that malnourished patients are 2 or 3 times more likely to experience post surgical complications, increased mortality, and prolonged hospital stay that can be 90% longer than their counterparts with an estimated 35%-75% increase in cost [6,7]. In one study, weight loss and a low BMI (Body Mass Index), FFM (Fat-Free Body Mass), or SMM (Skeletal Muscle Mass) alone was found to increase the risk of mortality by 1.45, 2.42, and 3.2 times respectively [8]. In another study on hospitalized patients, a group of 21% of patients who had a daily intake of <50% of recommended, were exposed to a risk of mortality 8 times higher than their counterparts [9] – For more information on Nutrition in Geriatrics and associated common diseases visit **Appendices A & B**.

In order to maintain a high standard level of optimum health in our aging population, the notion of healthy nutritional status must be assessed in elderly population so as to address the problems and needs of elderly people. Presently in Riyadh, there are little known cohort studies that address our geriatric community, and elderly care centres are fighting the exhausting turmoils of aging on their own with no common collaborations. It is seen as normal, whilst literature review has shown that successful aging can be one of life's wondrous gifts to one's body, spirit, mind and family which will ultimately decrease costs and economical burden in hospitals. Thus, this study is designed to determine three main key points. Firstly, a sample assessment size of nutritional status in elderly patients. Secondly, reviewing literature to establish the co-existence of nutritional status with common diseases of aging, and thirdly, addressing geriatric community at large, by a) increasing awareness about successful aging and b) addressing the Elderly Care Centre in Riyadh, and discussing its roles and duties in helping our geriatric community.

Methods:

Subjects: This study was carried out on elderly people, residing in in-patients ward at KFSH & RC, Riyadh, Saudi Arabia. Subjects who were over 60 years of age were selected in the study, and were picked up randomly in Cardiac and GI wards. Subjects have been residing in the hospital for at least a day, and some were about to be discharged. The analysis was heavily retrospective on their nutritional status and food intake before their admission to the hospital for at least 3 months, without addressing the current nutritional status at the hospital now. A total of 17 patients (male/female: 1.43) were included in the study.

Methods: in this study, data of 17 elderly patients were collected using Mini Nutritional Assessment (MNA) Questionnaire.

The Mini Nutritional Assessment (MNA) has recently been designed and validated to provide a single, rapid assessment of nutritional status in elderly patients in outpatient clinics, hospitals, and nursing homes. It has been translated into several languages and validated in many clinics around the world. The MNA test is composed of simple measurements and brief questions that can be completed in about 10 min. Discriminate analysis was used to compare the findings of the MNA with the nutritional status determined by physicians, using the standard extensive nutritional assessment including complete anthropometric, clinical biochemistry, and dietary parameters. The sum of the MNA score distinguishes between elderly patients with: 1) adequate nutritional status, MNA > or − 24; 2) protein-calorie malnutrition, MNA < 17; 3) at risk of malnutrition, MNA between 17 and 23.5. With this scoring, sensitivity was found to be 96%, specificity 98%, and predictive value 97%. The MNA scale was also found to be predictive of mortality and hospital cost. Most importantly, it is possible to identify people at risk for malnutrition, scores between 17 and 23.5, before severe changes in weight or albumin levels occur. These individuals are more likely to have a decrease in caloric intake that can be easily corrected by nutritional intervention [1][29].

It compromised four assessment domains [30] (a copy provided in appendices):

i.	Anthropometric Assessment	✓ BMI – Body Mass Index ✓ Mid upper arm and calf circumferences ✓ Weight loss during the past 3 months
ii.	Global Evaluation	✓ Accommodation type – living independently or in nursing home, ✓ Taking more than 3 prescription drugs, ✓ Psychological stress or acute disease in the past 3 months, ✓ Mobility, ✓ Neuropsychological problems, ✓ Pressure sores or skin ulcers
iii.	Dietetic Assessment	✓ Quantity and quality of eating meals, ✓ Loss of appetite, ✓ Digestive problems, chewing or swallowing difficulties causing decline in patient food intake,

		✓ Beverages consumed per day, ✓ Mode of feeding
iv.	Subjective Assessment	✓ Does patient consider having any nutritional problems? ✓ How would the patient consider his health status in comparison with other people of the same age?

For this study, the MNA questionnaire was interpreted promptly into Arabic by a researcher and filled up as he questioned patients. It took on average 7 minutes for each patient. Note, certain criteria were not met, such as BMI, Mid Arm/Calf Diameter, because it was not reliable in a retrospective scenario like this, and a limiting factor was: most patients were unable to get out of bed or were medically unsuitable to conduct to be measured.

<u>Field Study:</u>

A visit was conducted to Riyadh Elderly Care Centre for Males, where we interviewed several people there. It was under the supervision of the Ministry of Social affairs. We have found out that it was first established around 60 years ago, and they had a lot of goal enhancements over the past relatively 20 years. They focused on several domains of CARE, in Summary:

1) A social care domain which focuses on establishing a steady social residency state for the elderly and helping them in the phase they are going through.
2) A health care domain which focuses on the optimization of the well-being of the resident, and aims to prevent him or her from the common geriatric diseases and thus any health deterioration thereof.
3) A Cultural and Religious domain which aims on strengthening the beliefs of the resident by arranging discussion groups, and encouragement to do the spiritual activities mandated by their religious beliefs, i.e., Islamic Beliefs.
4) Professional & Handcrafts domain which aims to provide residents with training courses on certain hobbies dependent on their preference and acceptance.
5) Psychological domain that aims to monitor residential cases with regular routine follow-ups so as to ascertain their psychological soundness and sensibility and lessen their worries in order to help them better accept themselves, others, and surroundings. Regular discussion activities are arranged and encouraged among residents.

<u>Pamphlet:</u>

A pamphlet was created using various resources, mentioned in bibliography, which focuses on successful aging and identifying common findings about age disease. Its goal is to increase awareness and encourage optimizing our lives.

Results:

A total of 20 patients (male/female: 1:1) were included in the study. Mean age of the patients was 69.6 (SD 9.19 , R 60-87). According to different age groups, 65% (n=13) were youngest old (60-70 yrs), 20% (n=4) were middle old (70-80), and 15% (n=3) were oldest old (>80). MNA showed poor nutritional status in 85% of patients (30% at risk (MNA score <=23), 55% malnutrition (MNA score<17)) When different age groups were taken into consideration, 84.6% of youngest old had poor nutrition (30.8% at risk, 53.8% malnutrition), all middle old had poor nutrition (25% at risk, 75% malnutrition), and 75% of oldest old had poor nutrition (33.3% at risk, 66.7% malnutrition). The difference in the prevalence of malnutrition between age groups is not statistically significant. When gender was taken into consideration, 80% of males were found to have poor nutrition (60% at risk, 20% malnutrition) and 90% of females were found to have malnutrition with none in the "at risk" category. The prevalence of malnutrition in females is significantly higher than that in males (P<0.05). [For Data Tabulation visit **Appendix C**]

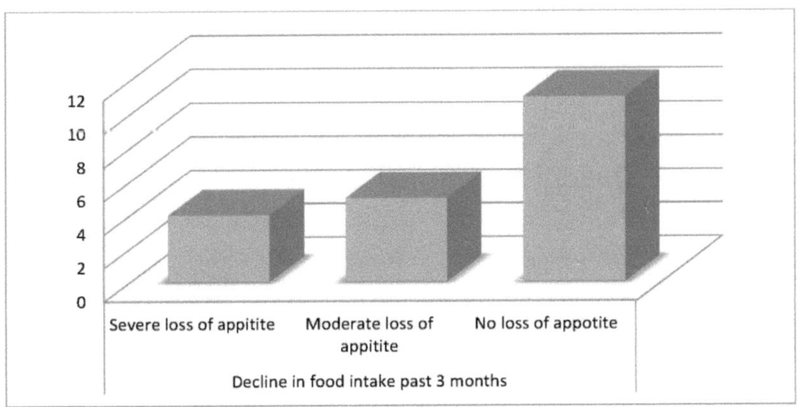

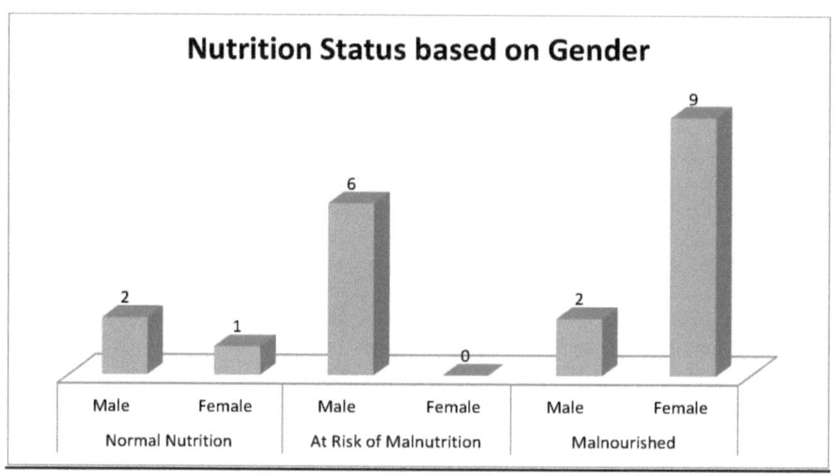

Nutrition Status based on Gender
2
1
6
0
2
9
Male
Female
Male
Female
Male
Female
Normal Nutrition
At Risk of Malnutrition
Malnourished

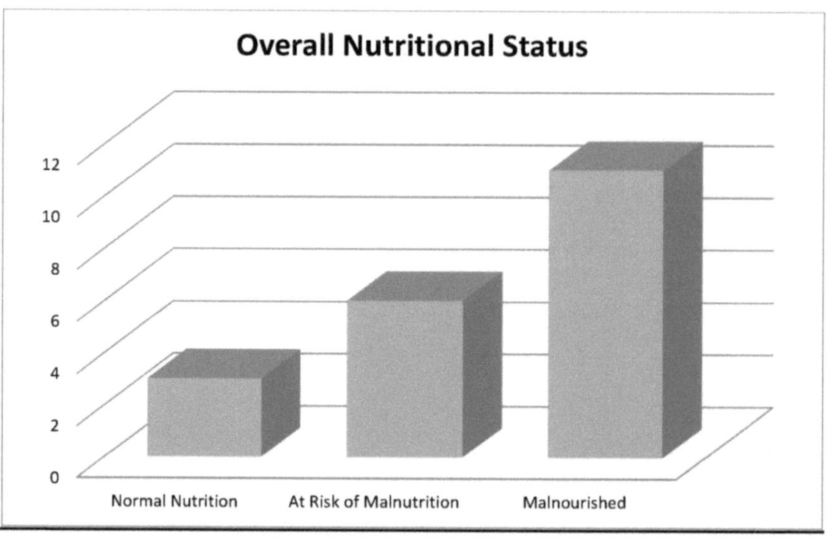

Overall Nutritional Status
12
10
8
6
4
2
0
Normal Nutrition
At Risk of Malnutrition
Malnourished

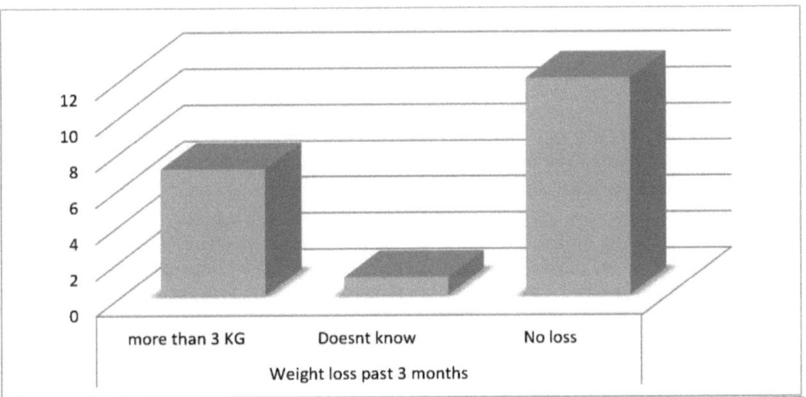

Literature review:

Due to decrease in fertility rates and mortality rates worldwide, a radical change is taking place in the population composition with a transition towards more old people (older than 60). This change will also shift the epidemiology more towards chronic non-infective diseases. The percentage of the world's population that is aged above 60 was 11% in 2009 and is expected to double by 2050. This percentage is much higher in developed countries at 21% and is expected to reach 33% by 2050. Less developed countries have it at 5-8%, and it's projected to reach 11-20%. The ratio of working to non-working population is expected to drop sharply, posing a future economic challenge, with an aging population requiring more health care costs. The oldest old group (>80), which presents the largest disease burden, is expected to increase at a faster rate as the life expectancy keeps on improving. Life expectancy's increase has been steady in the past with technological, medical, and life style quality advancements. Although there's a theoretical limit at which extending human life becomes unviable, it has not been reached yet and is not expected to be in the near future. The proportion of deaths due to infective diseases have fallen in favor of an increase in non-infective diseases such as diabetes, cancer, and cardiovascular diseases and is expected to continue rising. In developed countries, the top cause of disease burden was non-infective diseases at 85% followed by injuries and infective/nutritional/maternal conditions at 9% and 6% respectively in 2002. Non-infective diseases are projected to keep the lead in 2030 with 89%. [26,27]

Malnutrition is accompanied with other geriatric conditions and diseases. A study on 413 elderly patients showed that poorly nourished patients had more chronic diseases and geriatric syndromes [10], all of which can be traced back to frailty in elders, which is inter-correlated with malnutrition.

Anorexia:

Appetite and food intake steadily decreases with age, even when an elderly patient does not have a chronic condition. Studies show that the prevalence of underweight increases with each decade after 60. A decrease in the responsiveness of the physiological process behind compensation for energy demands by an increase in satiability is thought to be the culprit. This reduced ability to regulate appetite may contribute to anorexia in older patients [11-13].

Protein energy/Sarcopenia:

Decreased protein intake, also known as Protein Energy Undernutrition (PEU), is associated with an increase in protein catabolism throughout the body which leads to loss of body weight, loss of appetite (cachexia), and muscle wasting (sarcopenia) [14]. Prevalence of PEU is estimated to be 12%-85% [15]. The exact mechanism behind this catabolic state is thought to be triggered via disease-induced inflammatory processes. With an increase in catabolism, pressure ulcers can develop and wound healing is delayed [16,17]. Sarcopenia can develop into a disability. A study showed that patients with severe sarcopenia have a 79% more chance of developing disability [18].

Dysphagia/Malabsorption:

The GI tract shows little changes in healthy aging. However, esophageal motility may be affected with a marked decrease in the efficiency of peristalsis [19]. Certain medications that elderly patients with chronic conditions may be receiving can affect the absorption of some vitamins and minerals [20,21]. Poor oral hygiene or missing teeth can cause reduced food intake [22]. The prevalence of dysphagia is estimated to be 7-28% [23]. Problems can occur during chewing swallowing or food passing down through the esophagus. Intestinal absorption might be reduced. In a study, 21% of elderly patients at risk or malnourished had increased mucosal permeability leading to malabsorption and 4% had celiac disease [24].

Cognitive function:

Deteriorating cognitive function due to aging and malnutrition can mutually affect each other. A study on individuals aged 60 and above living in special housing in Sweden found a relationship between cognitive ability and nutrition. Older individuals with moderate to severe cognitive dysfunction had a higher risk of being malnourished regardless of housing and living arrangement [25].

Roles and Duties of Riyadh Eldery Care Centers:

Knowing their domains of CARE does not necessarily provide a nice overview for the reader about what their day-to-day challenges and responsibilities are in this Elderly Care Centre. So we decided to interview certain people and ask them about their jobs and responsibilities, and hereby sets of duties and roles for the structural hierarchy, from Administrator to Chauffer, are provided below:

1) Administrator's job is not simple for it involves so many guidelines and sets of principals that are vital to the overall function of the organization and to summarize in bullet points may not necessarily provide a good overview for what he is doing specifically, so as a general rule we can say an administrator's job involves three main categories. First: Vision; to never lose sight of the actual establishment itself and its utmost humble and humane goal: That we all are children of our elderly geriatric fathers and mothers, thus, working hard to please them and make sure that their necessities are met socially, psychologically, medically, and overall generally. Second: Routine work; making sure that all admission requests for the house are meeting a common standardized criteria for admission, and establishing the various responsibilities for other directors and jobs in the house, whilst making sure the day to day challenges of each director, assistant or any other

employee are being met on a high standard quality of care and discipline. Third: Planning; involving different committees of the organization to provide the best standard of care for any individual whilst making sure that research and studies that lead to the enhancement of the house are suggested to responsible authorities. Also, establishing the firm's communication and networking to other corresponding areas of interests, whether it is the ministries, hospitals or the like.

2) Social Service Director: To be fully responsible for the social aspects of residential cases which ranges from evaluating them for admission, to orienting them around the house, and supervising the cultural and social programs established. However, this is always rendered valuable when the resident's social care is met with care, necessitating for him to open up social profile for each resident, and making sure the social lives of the residents are enforced with better relationship skills between themselves, other residents in the house, and their families or relatives.

3) Psychologist's job entitles so many psychological maneuvers and tricks in order to enhance the inner lives of residents. Primarily, newly admitted cases are assigned to clinical assessment by the psychologists, and followed by routine check-ups. Nevertheless, opening psychological profiles for each resident is mandatory so as to continually conduct post-hospital evaluations and write reports of any newly abnormal psychiatric cases; discussing any cases with the medical team is also part of his job, as well as finalizing any further or necessary work for any re-admission to psychiatry wards in hospitals.

4) Supervisor's job is probably one of the toughest and most essential jobs in the house for it mandates a high set of vigilance and observation skills for residents on 24/7 basis. His job also entitles doing required tasks by other directors (e.g., social director), and he is fully responsible for any case of misconducts or smuggling in the house. He plays a major role in assessing personal care and hygiene among residents too.

5) The Doctor: This white collar position is redeemed very important clinically for assessing overall general health of residential cases. It mainly revolves around management and treatment of patients into two domains: Treating by prevention, and treating by intervention. The former involves annual vaccinations for elderly residents, taking notes, recording their cases and discussing them with the medical team when required, educating the residents on their health, and writing the reports about their medical cases. Whilst, the latter involves diagnosis, management, and treatment work-up of any newly emerging cases of disease in the house; also, evaluation for any required isolation for any case in wide-spread endemic diseases.

6) The Nursing Staff responsibilities are always demanding. It focuses on the follow-ups of patient's compliance with medication and treatment. It involves so many responsibilities summarized in the following categories: a) Health category: it is concerned with patients' health status by follow-ups for patients' continual medicinal treatments prescribed by doctors, and monitoring patients' compliance. It also entitles the follow-ups for weight management programs, observing dietary intakes, and observing their normal eating habits during meals b) Personal Hygiene & Organization category: it involves scheduling of personal hygiene sessions with social director for residents, and supervision of their overall obedience. c) Emergency Category: it involves first aid procedures whenever necessary, helping doctors in acute emergent cases, and accompanying patients to hospitals whenever necessary.

7) Physiotherapist: His job involves conducting physiotherapy programs according to resident's health status, increasing their awareness to certain activities and its relation to physiotherapy, and making sure it abides to safety and security guidelines

8) Registered dietician's role involves dietary intake of residents, from their caloric intake to the ingredients set up in the kitchen and its expiration date. It generally involves observation, assessment of their dietary life style, and studying resident's complaints about food.

9) Others, which involve the following:
    a. Activity director where he participates in the establishment of various activity programs.
    b. Personnel Coordinator where his job involves managerial procedures.
    c. Housekeeping services who are four:
        i. Cooking Chef, whose main job is, obviously, cooking high quality food.
        ii. Laundryman, who washes clothes of the residents and ensure they look neat.
        iii. Guard, whose tasks are known to the reader.
        iv. Chauffer whose job is to drive to distant places and drive for residents or other tasks assigned by administrator.

Pamphlet:

Below is the pamphlet produced, it was translated into Arabic and emailed to University Campus:

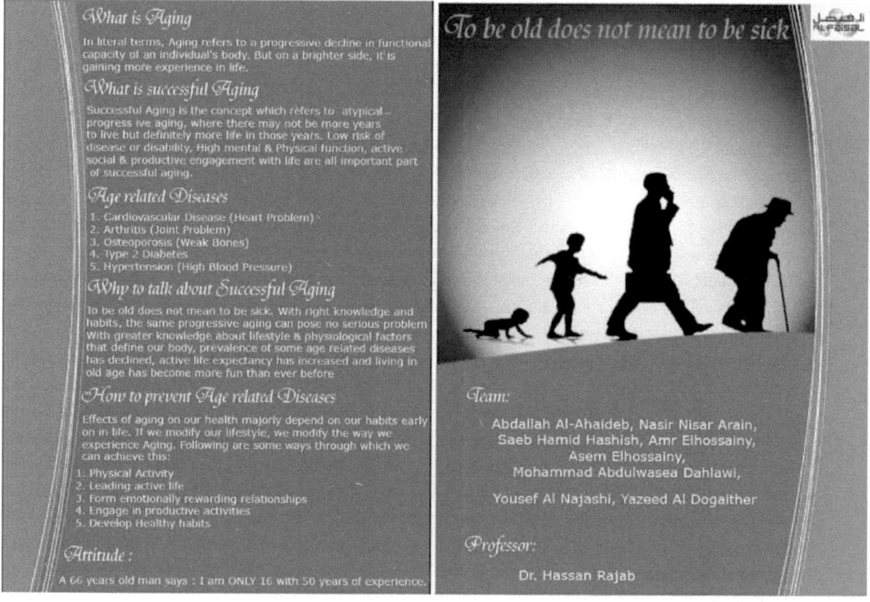

Discussion:

The elderly are becoming a fast growing section of the population with malnutrition emerging as an important problem contributing to morbidity and mortality. Problems contributing to malnutrition include gastrointestinal and endocrine disorders, loss of taste and smell acuity, and decreased appetite.

The prevalence of malnutrition found in this study is higher than in other studies in the literature, almost double that of a Belgian study [28], and 20% higher than in a multinational study done in 12 countries [1]. Levels of malnutrition did not differ significantly between gender and age groups.

A major limitation of the study is the small sample size and results may not be generalized for elderly of the region. Another limitation is that all patients interviewed were in a hospital setting and other settings are not represented. No data on chronic conditions and diseases were collected from the patients. Future studies with a larger sample size spanning diverse settings may be needed to more accurately assess malnutrition of the elderly in the region and correlation with common conditions among the elderly might prove useful.

Conclusion:

The study confirms and iterates the idea that there is co-existence between malnutrition and other diseases, and that malnutrition is a common problem among elderly. Though, due to the small sample size, this cannot be generalized for all population. Aging is not to be addressed lightly and in a single, confined format approach but rather a diversely comprehensive one demonstrating challenges in a number of public policy areas. Nor is aging confined only to older populations in KSA for example; it is an issue that affects all generations as the demographic structure of the population changes. All hierarchies will need to cooperate to respond to the challenges of an aging society in an inclusive manner, for the overall benefit of people.

References:

1- Kaiser M, Bauer J, Sieber C, et al. Frequency of Malnutrition in Older Adults: A Multinational Perspective Using the Mini Nutritional Assessment. Journal Of The American Geriatrics Society [serial online]. September 2010;58(9):1734-1738. Available from: Academic Search Premier, Ipswich, MA. Accessed January 3, 2012

2-Russell CA, Elia M, on behalf of BAPEN and collaborators: Nutrition screening survey in the UK in 2007. Nutrition screening survey and audit of adults on admission to hospitals, care homes and mental health units. <http://www.bapen.org.uk/pdfs/nsw/nsw07_report.pdf> BAPEN 2007.

3-Gang Wang, Ying Wan, Qi Cheng, et al. Malnutrition and associated factors in Chinese patients with Parkinson's disease: Results from a pilot investigation. Parkinsonism and Related Disorders 16 (2010) 119-123

4- D. Hrnciarikova, Z. Zadak, Malnutrition in seniors in the Czech Republic, European Geriatric Medicine, Volume 2, Issue 2, April 2011, Pages 111-114, ISSN 1878-7649, 10.1016/j.eurger.2011.01.014. (http://www.sciencedirect.com/science/article/pii/S1878764911000234)

5- Bruno J. Vellas, William C. Hunt, Linda J. Romero, Kathleen M. Koehler, Richard N. Baumgartner, Philip J. Garry. Changes in nutritional status and patterns of morbidity among free-living elderly persons: A 10-year longitudinal study. Nutrition - June 1997 (Vol. 13, Issue 6, Pages 515-519)

6- REGULA MÜHLETHALER, ANDREAS E. STUCK, CHRISTOPH E. MINDER, and BRIGITTE M. FREY. The Prognostic Significance of Protein-energy Malnutrition in Geriatric Patients Age Ageing (1995) 24(3): 193-197 doi:10.1093/ageing/24.3.193

7- Norman DC. Nutrition and failure to thrive. In: Osterweil D, Reuben DB, Rozzini R, Rubenstain LZ, Trabucchi M editor. New frontiers in geriatric medicine. Brescia: Kendal; 1992

8- Simcha Kimyagarov, Raisa Klid, Shalom Levenkrohn, Yudit Fleissig, Bella Kopel, Marina Arad, Abraham Adunsky, Body mass index (BMI), body composition and mortality of nursing home elderly residents, Archives of Gerontology and Geriatrics, Volume 51, Issue 2, September-October 2010, Pages 227-230, ISSN 0167-4943, 10.1016/j.archger.2009.10.013.

(http://www.sciencedirect.com/science/article/pii/S0167494309002684)

9- D.H. Sullivan, S. Sun and R.C. Walls, Protein-energy undernutrition among previous termelderlynext term hospitalized patients: a prospective study. JAMA, 281 (1999), pp. 2013–2019.

10- Bulent Saka, Omer Kaya, Gulistan Bahat Ozturk, Nilgun Erten, M. Akif Karan, Malnutrition in the elderly and its relationship with other geriatric syndromes, Clinical Nutrition, Volume 29, Issue 6, December 2010, Pages 745-748, ISSN 0261-5614, 10.1016/j.clnu.2010.04.006.

(http://www.sciencedirect.com/science/article/pii/S0261561410000841)

11- I.M. Chapman, The anorexia of aging. Clin Geriatr Med, 23 (2007), pp. 735–756.

12- Z. Kmiec, Central regulation of food intake in ageing. J Physiol Pharmacol, 57 Suppl. 6 (2006), pp. 7–16.

13- S.B. Roberts, P. Fuss and M.B. Heyman, et al. Control of food intake in older men. JAMA, 272 (1994), pp. 1601–1606.

14- D.R. Thomas, Guidelines for the use of orexigenic drugs in long-term care. Nutr Clin Pract, 21 (2006), pp. 82–87.

15- A.J. Silver, J.E. Morley and L.S. Strome, et al. Nutritional status in an academic nursing home. J Am Geriatr Soc, 36 (1988), pp. 487–491.

16- Langer G, Schloemer G, Knerr A, Kuss O, Behrens J. Nutritional interventions for preventing and treating pressure ulcers (Cochrane Review). The Cochrane Library. 2004;1.

17- Gunningberg L. Prevention of pressure ulcers in patients with hip fractures: Definition, measurement and improvement of the quality of care. 2000. <http://publications.uu.se/uu/fulltext/nbn_se_uu_diva-559.pdf>

18- Janssen I. Influence of Sarcopenia on the Development of Physical Disability: The Cardiovascular Health Study. Journal Of The American Geriatrics Society [serial online]. January 2006;54(1):56-62. Available from: Academic Search Premier, Ipswich, MA. Accessed January 3, 2012.

19- J.E. Morley, The aging gut: physiology. Clin Geriatr Med, 23 (2007), pp. 757–767.

20- T.S. Dharmarajan, M.R. Kanagala and P. Murakonda, et al. Do acid-lowering agents affect vitamin B12 status in older adults?. J Am Med Dir Assoc, 9 (2008), pp. 162–167.

21- S.E. Gulmez, A. Holm and H. Frederiksen, et al. Use of proton pump inhibitors and the risk of community-acquired pneumonia: a population-based case-control study. Arch Intern Med, 167 (2007), pp. 950–955.

22- N. Dion, J.L. Cotart and M. Rabilloud, Correction of nutrition test errors for more accurate quantification of the link between dental health and previous termmalnutritionnext term. Nutrition, 23 (2007), pp. 301–307

23- K. Kawashima, Y. Motohashi and I. Fujishima, Prevalence of dysphagia among community-dwelling previous termelderlynext term individuals as estimated using a questionnaire for dysphagia screening. Dysphagia, 19 (2004), pp. 266–271.

24- Terry Bolin, Marian Bare, Gideon Caplan, Suzie Daniells, Margaret Holyday. Malabsorption may contribute to malnutrition in the elderly. Nutrition - July 2010 (Vol. 26, Issue 7, Pages 852-853, DOI: 10.1016/j.nut.2009.11.016)

25- Cecilia Fagerström, Roger Palmqvist, Johanna Carlsson, Ylva Hellström. Malnutrition and cognitive impairment among people 60 years of age and above living in regular housing and in special housing in Sweden: A population-based cohort study. International Journal of Nursing Studies - July 2011 (Vol. 48, Issue 7, Pages 863-871, DOI: 10.1016/j.ijnurstu.2011.01.007)

26- World Health Organization: The Global Burden of Disease: 2004 Update. © World Health Organization 2008. Accessed Jan 3, 2012 at http://www.who.int/healthinfo/global_burden_disease/2004_report_update/en/index.html.

27- Mathers CD, Loncar D: Projections of global mortality and burden of disease from 2002 to 2030. PLoS Med 3:e442, 2006

28- Katrien Vanderwee, Els Clays, Ilse Bocquaert, Micheline Gobert, Bert Folens, Tom Defloor. Malnutrition and associated factors in elderly hospital patients: A Belgian cross-sectional, multi-centre study. Clinical Nutrition - August 2010 (Vol. 29, Issue 4, Pages 469-476, DOI: 10.1016/j.clnu.2009.12.013)

29- Vellas B, Guigoz Y, Garry PJ, et al. The Mini Nutritional Assessment (MNA) and its use in grading the nutritional state of elderly patients. Nutrition 1999; 15:116.)

30- Saidlou, et al. Assessment of The Nutritional Status and Affecting Factors of Elderly People Living at Six Nursing Home in Urimia, Iran. International Journal of Academic Research Vol. 3. No.1. January, 2011. < http://www.ijar.lit.az/pdf/9/2011(1-28).pdf>

31- MNA Mini nutritional assessment. Available at www.mna-elderly.com. (Accessed December 28th, 2011).

Bibliography for Pamphlet:

1. Aging successfully - Social & Community aspects of Aging (Rodney M.Coe, John E.Morley, Nina Tumosa)
2. Successful Aging - George Burns

Appendix:

Nutrition in Geriatric Population & Prevalent Diseases:

A sturdy process indeed it is dealing with nutrition in the elderly; most of the times they are poorly compliant with dietary recommendations while having a low appetite adding to the decreased body mineralization. Elderly patients tend to like certain food types, which accompanies weight gain with all the calories it holds without benefit to their frail body, meanwhile, since they are old with declined body metabolism, doctors resort to food restriction. Food restriction can have a counterproductive effect not just on the long term physical health of the patient (weight loss), but also on their psychological health. Thus, accurate planning of their nutritional health is needed, making life harder for nutritionists to balance the body <u>needs</u> with the body <u>desires</u>. Elders, who may not tolerate much chewing or eating, being it a pathological state or an energy consuming process, need a plan that integrates easy and maximum nutritional gain (e.g., liquid condensed supplements or pills). As consequences for aging, the patient faces physiological losses of minerals like protein, calcium/vitamin D, vitamin b12, as well as decreased contractions and thirst sensation, all leading to problems like muscle degeneration, bone mass reduction and fracture risk, anemia, constipation, and dehydration respectively, which accounts for almost all aging and frailty problems. A food pyramid needs to take all that into account, balancing all circumstances in the process.

Modified MyPyramid for Older Adults

Source: http://www.dentalcare.com/media/en-US/education/ce301/ce301.pdf

Biologically, systemic consequences of aging are widespread but can be clustered into four main domains or processes: a) body composition, b) Balance between energy availability and energy demand, c) Signaling networks that maintain homeostasis, d) Neurodegeneration. Those 4 consequences make the basis for most symptoms and heath declivity leading to frailty. Frailty itself has 4 main consequences for clinical practice: Low resistance to stress, Multiple diseases lead to frailty, and vice versa (Comorbidity & Polypharmacy), Disability and Impaired Recovery from Acute-Onset Disability, Geriatric Syndromes: Geriatric syndromes include: a) Urinary incontinence, b)Dementia. C) Falls (which can really affect the patient's life tremendously if prevented).

Urinary Incontinence is the involuntary loss of urine. It is present in about 30% of people more than 65 years old, and more than 50% of nursing home residents. Common complications of Urinary Incontinence include sleep deprivation, embarrassment, social withdrawal, depression, limited activity. Anatomical and physiological changes in the elderly that can cause Urinary Incontinence include Benign Prostatic Hyperplasia in males, Atrophic Vaginitis in postmenopausal women, Atrophic Urethritis in males, detrusor muscle overactivity in the bladder, decreased ability to postpone urination, and decreased total bladder capacity. Risk factors for Urinary Incontinence include diabetes, Parkinson's disease, moderate to severe dementias, strokes, and obesity.

Dementia is the progressive impairment and eventually loss of memory and cognitive functions of the brain. Dementia is the most common cause of disability in elderly patients. The most common cause of dementia is Alzheimer's disease. Dementia is a multi-functional disorder in which the cognitive functions such as memory, learning, reasoning, attention, language, recognition, problem solving, and emotion are affected. In the late stage of dementia, patients have difficulty in recognizing time, place, and person. Signs of Dementia may include tremors, rigidity, bradykinesia, difficult initiation of movement, and feet dragging. Delusions of persecution, agitation, and aggression also often accompany dementia. The patient's behavior is disorganized. Dementia patients have difficulties in controlling their temper and emotion when they are put in stressful situations or circumstances beyond their abilities.

Further Readings:

- Ch70 World Demography of Aging - Harrison's Principles of Internal Medicine. New York. McGraw Hills (18th ed)

-Ch71 The Biology of Aging - Harrison's Principles of Internal Medicine. New York. McGraw Hills (18th ed)

-Ch 72 Clinical Problems of Aging - - Harrison's Principles of Internal Medicine. New York. McGraw Hills (18th ed)

Data tabulation, mean, and standard deviation are used as descriptive statistics. Statistical variation in categorical variables is analyzed using the chi-square test.

Table – 1:

Gender/ Status	Normal	At Risk	Malnutrition	Total
Male	2	6	2	10
Female	1	0	9	10
Total	3	6	11	20

Expected: Contingency Table

Gender/Status	Normal	At Risk	Malnutrition
Male	1.50	3.00	5.50
Female	1.50	3.00	5.50

- ☒ Chi-square = 10.8
- ☒ Degrees of Freedom = 2.00
- ☒ Probability = 0.0045

Table – 2:

Gender/status	Nomral	At Risk	Malnutrition	Total
Young Old	2	4	8	13
Middle Old	0	1	3	4
Oldest Old	1	1	1	3
Total	3	6	11	20

Expected: Contingency Table

Age/Status	Normal	At Risk	Malnutrition
Young Old	1.95	3.90	7.15
Middle Old	0.60	1.20	2.20
Oldest Old	0.45	0.90	1.65

- ☒ Chi-square = 1.87
- ☒ Degrees of Freedom = 4.00
- ☒ Probability = 0.760

YOUR KNOWLEDGE HAS VALUE

- We will publish your bachelor's and
  master's thesis, essays and papers

- Your own eBook and book -
  sold worldwide in all relevant shops

- Earn money with each sale

Upload your text at www.GRIN.com
and publish for free